A Wine Tourist's Guide:

Visiting Tasting Rooms

A Wine Tourist's Guide:
Visiting Tasting Rooms

by

Terry & Kathy Sullivan

INFINITY
PUBLISHING

ISBN 978-0-7414-9801-4

Printed in the United States of America

Published September 2013

Book Website
http://awinetouristsguide.winetrailtraveler.com

INFINITY PUBLISHING
1094 New DeHaven Street, Suite 100
West Conshohocken, PA 19428-2713
Toll-free (877) BUY BOOK
Local Phone (610) 941-9999
Fax (610) 941-9959
Info@buybooksontheweb.com
www.buybooksontheweb.com

Dedication

This book is dedicated with love to our children: Ryan, Erin, Meghan, Kevin, and those special people in their lives - Gwen, Gerard, Clara and Andy.

We love you all!

Contents

Introduction

It is a nice weekend afternoon and you decide to get away and visit a winery or vineyard tasting room. Where do you go and what do you do once you get there? This guide gives some practical suggestions about what to do when visiting a tasting room.

Visiting a tasting room is entirely different than going to a bar and ordering a glass of wine. A winery and vineyard have a certain romance associated with the setting. Even wine tasting at a small down-to-earth mom and pop tasting room has a certain elegance. Expectations in a tasting room setting are different than expectations in a bar setting.

During the last seven years we have visited and written about 1,000 tasting rooms on three continents. We share what we have learned in this book. Discover the expectations and etiquette when visiting tasting rooms. Learn the process of tasting wines and about some of the world's international grapes. Become familiar with myths and misconceptions wine enthusiasts share with tasting room staff. Explore ways to find winery and vineyard tasting rooms to visit throughout the year.

Wine encompasses a whole world that includes history, art, biology, chemistry, and education. Enjoying wine can be done simply with a good book or with friends. However wine is much more.

Chapter 1:
Tasting Room Expectations

What do you expect when you open the door of a tasting room? We have spoken to several people who have told us they would not visit a tasting room because they did not know what to expect, and that made them feel uncomfortable. They commented that they would not know what to do. Other wine enthusiasts have no concerns about walking into a winery tasting room. We were apprehensive the first time we walked into a winery tasting room. We were not sure what to expect and the setting was unknown.

Most tasting rooms are similar in the way they operate. Some of the most pleasant attributes of visiting wineries are the friendliness of the staff and the winemaker's passion for winemaking. When you visit a winery you can expect people to be friendly. Some tasting rooms will have a greeter near the entrance who will direct you to a tasting counter. If not, look for the tasting counter, usually a bar type setup with a server behind it. You walk up to the tasting counter and request a tasting.

Some wineries charge for tastings, and others do not. Occasionally, a souvenir glass is included when you pay a tasting fee. Whether you pay a fee or the wine tasting is free, you will be given a list of wines to taste. Some wineries will allow you to try everything on the list; others will allow you to select five or six wines to taste. If you are traveling with a companion and you are both tasting, you can both taste five different wines and share. Do keep in mind the amount of wine that you consume. If a winery permits you to taste everything, you may want to only visit one or two wineries that day. Drink plenty of water before, during and after a wine tasting.

Let the winery staff guide your tasting. We have experienced consultants who pour the wine and tell us what we are about to smell and taste. Others will ask us what we observe about the wines. However, the latter is less frequent. There is a certain risk that winery staff take if they ask your opinion. At one winery, a staff member had a look of disappointment when she asked Terry what he thought a wine tasted like and he replied, "leather." The person conducting the tasting thought it tasted like coffee. If you feel intimidated when asked about a wine, just say, "It tastes like a lot of hard work went into its making."

Wine tasting at DuCard Vineyards in Etlan, Virginia

If you go to a tasting room on a less busy day (usually in the middle of the week), you will enjoy the opportunity to talk about the wine, winery and vineyards. Most tasting room staff members are knowledgeable about the wines they are pouring. A mere one percent of the wineries we have visited had a staff member who admitted that he just helps on the weekend and pours the wine. He told us he did not know anything about the wines he was pouring. This is a very rare occurrence as most winemakers and owners train their staff.

Most of the time, you will stand at the tasting counter. Some tasting rooms have bar stools next to the tasting counter. This is a very welcome addition if you are tasting at your third winery of the day. Some wineries use tables for wine tastings. At McGregor Vineyard and Winery along Keuka Lake in New York, you will sit at a table. The tasting room staff bring a plate of bite-sized food and discuss the wine list with you. We also had a "sit down and be served" experience at several wineries in California.

Sit down wine tasting at Shafer Vineyards in Napa, California.

The "sit down at a table" theme continued at Schramsberg in California. After a tour of the caves at Schramsberg, the tour group gathered in a small dining room that had three tables set for a group of people. Our tour filled two of the tables. The wine consultant talked about the sparkling wines and then served us at the table. This afforded a great opportunity to talk about the wine with the other people at our table. This intimate approach also occurred after group tours at Shafer, Caymus, Roy, Pine Ridge, Gloria Ferrer and Quintessa in California.

While at the tasting counter, you can make some decisions. One is what to do with the wine you are tasting. Your options include swallowing it or spitting it out into a dump bucket. Many people just swallow since they are not comfortable spitting a wine. You can ask for a small cup that makes spitting much easier. You can also take the opportunity to practice spitting while taking a shower. With practice, you should be able to hit the drain. Spitting becomes much easier when spitting into a dump bucket.

Occasionally the winemaker will be on hand and may give a tour and conduct your tasting. One memorable tour we had was at The Lenz Winery on Long Island in New York. The winemaker had us taste his wines from the stainless steel tanks. He asked us what we thought, and more often than not, agreed with us. His tour proceeded to the barrel room, and with thief in hand, had us taste the wine from some of the barrels. Tasting wine from the barrels or tanks can give you an idea of what the wine is like at that moment in time.

One of our favorite tastings was during the tour at Del Dotto Winery in California. The tour led our group through the over one-hundred-year-old candlelit tunnels. Our guide stopped along the way and had us taste wine from the barrels. If we liked the wine, we took a ticket attached to the barrel. After the tour, we could order any of the wines we tasted. After ordering, the wine was bottled and shipped to the buyer.

Visiting winery tasting rooms is a great activity. Try to limit the number of tasting rooms you visit in one day. We try to visit two or three in a day. Tasting rooms are less crowded on the weekdays. Call ahead or check their times on the Internet.

Winemaker uses a thief to take a barrel sample at Megalomaniac in Vineland, Ontario, Canada.

Ask Questions

In a thirst for wine knowledge, more people are visiting winery and vineyard tasting rooms. One way to learn about wine is to taste and drink wine. Tasting rooms offer an informal experience that is ripe for asking questions. Try asking the same question at different tasting rooms and notice if the answers vary. For example, one may not know about French wines and ask if all Burgundies are red. Compare answers. One's wine knowledge will increase with the more questions one asks.

For those of you beginning on your wine journey, asking questions may be a challenge. People often do not ask questions because they are afraid of appearing ignorant. Ask questions. Ask questions not only of the tasting room staff, but start a dialogue with the other visitors next to you.

You can also research the answers to your questions. Do not always take for truth an answer. There are times we encounter ill-informed tasting room staff. While traveling along Virginia Route 29, a tasting room employee at a winery proudly stated to several guests that Virginia has almost as many wineries as California. To most people, Virginia's 250 plus wineries is not "almost as many" as California's more than 3,000 wineries. In Texas, a young tasting room staff member was giving a group tour of the vineyards and winery. He mentioned that American oak barrels are much larger than French oak barrels. This is not the case.

Some questions to ask while at a wine tasting:

Where were the grapes in this wine sourced?
Do not assume that the grapes came from a vineyard the winery owns. They might have, or they may have been shipped thousands of miles to the winery. A winery and a vineyard are two different businesses. Many wineries purchase grapes from growers both locally and a distance away.

What grape varieties are in this wine?
Do check the label. If it says "Chardonnay," then you can assume that most of the wine, 85% or more, is Chardonnay. However, if the label says "Big Dog" and it does not list the varietal grapes, ask away. Each country has its own set of regulations for wine labels.

What foods would you pair with this wine?
The traditional answer to this question was white wines go with seafood and poultry whereas red wines go with meats. However, tradition sometimes fades away. Now, many wine enthusiasts simply drink what they like with whatever they are eating. Further

investigation reveals that the right wine will enhance the food, and food can enhance the right wine. See what the winery suggests for food pairings, and give it a try.

How long can this wine age?
Many wines sold are meant to be opened and enjoyed immediately. Discover the answers you get from winery staff for this question about particular wines.

Why did the owners of this winery decide to start a winery?
After visiting one thousand wineries/vineyards, we still get unique answers to this question. One owner was with his family for a vacation on a flight from Pennsylvania to Florida. In the airline magazine, he read an article about planting a vineyard. During the vacation he read more about starting a vineyard. Returning home, he started to make plans to plant the vineyard. Now the winery and vineyard are in the capable hands of the second generation.

In addition to you asking questions, expect the person conducting your tasting to ask you questions. They often ask, "Where are you from?" One question to prepare an answer for is, "What wines do you like?" Our response to that question is, "We like any wine from Aglianico to Zweigelt." That response covers the alphabet. We are open to trying wines produced from varietal grapes that we have not tried before. Be honest with your answer. If you like California Cabernet Sauvignon, use that for your answer. If you like sweet wines, say you like sweet wines. If you like dry red wines, tell the staff member conducting the tasting that you like dry red wines.

As you ask questions about wine and collect answers, you will add knowledge to your wine journey. Together with tasting and drinking wines, this knowledge will continue to build and you will find yourself thirsting for even more wine knowledge. It is easy to start; just ask questions. Remember there are no ignorant questions, just ignorant answers.

After the Tasting

After your last wine is poured, there may be a hard sales pitch or no sales pitch to buy wine. While visiting one winery, our server took out an order form and asked which wines we wanted to buy while we were tasting our final wine. This tactic only happened once and made us uncomfortable. Most wineries do not have a sales pitch. Many of these wineries have the wines set up in racks near the tasting bar, and you can leisurely buy if you want. Occasionally, there may be more of a sales pitch from wineries that do not charge for a tasting. However, our experience indicates that most of these wineries do not have a sales pitch.

Should you buy any wine? Whether or not there is a sales pitch, do not buy the wine if you do not like it, even if the tasting is free. One of the purposes of a tasting is to introduce the public to the wine a winery produces. You cannot be forced to like them. On the other hand, if you like a wine, then you should consider purchasing it. Many wineries produce small numbers of cases and sell from the tasting room. They need customers who will purchase from the tasting room.

Winery Tours

A great experience before or after a tasting is an offer to take a tour of a winery and vineyard. Some tours are very short; you stand in one spot and listen to your guide point out the equipment and discuss the process. Other tours can last an hour or longer, and your guide will point out the equipment, discuss the process, ask you questions and answer your questions. You may be taken to the vineyard to learn about the grapes. The tour may take you into several rooms.

Many wineries offer tours on request, but they need to have the staffing to take you on a tour. Look at the winery's website. If they offer tours on request, call ahead and schedule a time. Occasionally, a winery will have regularly scheduled tours and will not require you to call to make an appointment. Tours are a wonderful way for you to learn about the winemaking process and learn about the wines a winery produces.

Goals for Your Visits

If this is your first visit to a winery, concentrate on what you like. Do you like dry reds, sweet wines, blush wines, dry whites, or others? Do you like wines that are 75 - 100 percent made from the same grape or do you like blends of wines from two or more grape varieties? Do you like fruit wines made entirely from the fruit or fruit wines made from grapes with the fruit flavoring added? When you leave the winery, identify the wines that you liked.

Take time to talk to your server. If you do not know the kinds of wines you like, let your server know so she may direct you to certain wines. If you are a beer drinker, she may suggest the type of wines you may want to try first. Do not rush deciding what you like in wines. Make several winery visits before you come to conclusions about what you like. Once you know what you like, start tasting those wines at different wineries. For example, if you like Chardonnay, try the Chardonnay at different wineries. There are so many variables in making a wine that two Chardonnay vineyards growing side by side may produce wines with different aromas and tastes.

Choosing Wineries to Visit

First, consider visiting wineries closest to where you live. Use the Internet and search for the wineries in your state. Chart a route to two or three of them. If there are more than three wineries near you, plan to visit the others at a different time. For the two or three wineries you plan to visit in a day, go to the winery's website and see if they have a tasting room and what times they are open to the public. Some wineries are open year round, while others open only on the weekends. Plan on spending at least a leisurely hour at each winery.

If you are familiar with wineries in your local area, branch out to the region, then on to other wine regions. There are wine regions all over the world waiting for your discovery. Do your research. Some wineries require appointments. Some may not have tasting rooms and are not open to the public.

Consider taking a wine tour offered by a wine tour operator. Again, you should do your research. Some tour operators only give tours in a certain region. Others give tours in several regions around the world. Some tour operators will provide guided tours. You will be picked up, usually at a hotel, and driven to wineries. The driver often talks to you about the wineries. Or, if the tour is for a larger group, a guide may accompany the group and talk about the wineries and wines you may be tasting.

Some tour operators will provide you with an itinerary, make appointments at the wineries and book a rental car and lodging. You then do the driving from one winery to another. This type of tour is usually arranged to meet your particular interests and needs.

Chapter 2: Tasting Room Etiquette

Many small, boutique wineries are family-run. When visiting a wine tasting room, consider yourself a guest. The owners and staff are proud of their facility and want everyone to enjoy their visit. Expectations in wine tasting rooms are different than at wine festivals or in bars. After visiting 1,000 wineries in the last seven years and talking with numerous winery staff and visitors, we have created a selection of ideas to keep in mind when visiting a winery tasting room. Simple rules of etiquette can help to make the experience enjoyable for all.

Subtle Sophistication

The atmosphere in a winery tasting room is one of subtle sophistication. While shorts and athletic shoes are acceptable, so are semi-dress clothes. Do not wear perfume or cologne the day you are visiting winery tasting rooms. The fragrances will interfere with a wine's aroma, not only for you, but also for those near you. Toothpaste may also interfer with your tasting; rinse with water.

Conversational Tones

In the sophistication of a winery tasting room, loud outside voices are not appropriate. Conversational tones are perfect and fit in well with talking about the wines you taste and meeting other wine enthusiasts who have common interests. Do take the time to talk with other visitors. It is interesting to meet others with a passion for wine. Some people are very new to wine or visiting wineries. Never belittle someone for what they think about a wine. Do not act or sound like a wine snob. There are always those who know less and more about wine than you.

Wine Limits

If you appear to be becoming intoxicated, the tasting room attendant is not permitted, by law, to serve you even if it is a tasting. Do not try to demand a tasting; drunken behavior is not acceptable. Wine is a beverage that should be appreciated as a social beverage and an accompaniment to food. Wineries deal with this behavior in different ways; most often you will be asked to leave.

Tasting Order

The wines for a tasting are presented in a particular order based upon the style of wine. Generally, white wines are served before red wines, and dry wines are served before sweet wines. Since the sweet wines are sweet, the sugar will make dry wines tasted after sweet wines seem bitter.

"I don't like that wine."

It is fine if you choose not to drink a particular wine. Gently cover your wine glass with your fingers to indicate you do not want to taste a particular wine. It is so much more subtle than declaring, "I don't like that wine." Not all wine drinkers like all wines, and wine hosts understand this.

After tasting a wine, visitors do not need to announce to everyone that they disliked the wine. Remember it is perfectly acceptable for you to dislike a wine. However, keep in mind that the person next to you may very well enjoy it.

Crowded Tasting Rooms

Tasting rooms can be crowded on weekends. Weekdays are generally slower, and wine hosts have more time to talk about the wines you taste. In either case, if the tasting room is busy, do not elbow your way to the tasting bar. When a tasting room is hectic, consider stepping back to discuss your wine and enjoy the tasting while talking with other like-minded people. This gives others room to step up for a tasting. On crowded days, some wineries will set up tasting tables or bars outside the tasting room. Give yourself plenty of time at a winery. Relax and enjoy the wine tasting and the ambiance of the tasting room. Many wineries have gift selections to browse while tasting wines. Take your time tasting and browse the offerings.

Tasting rooms can become crowded as at Te Mata Estate Winery Havelock North, New Zealand

The Dump Bucket

What should you do with the wine in your glass you have tasted? If you do not want the rest of the wine in your glass, you may pour it into a dump bucket. It is really okay. In addition, it is a good idea to spit your wine into the bucket or a plastic cup. You can then dump the contents from the cup into the dump bucket. Even though tastings are small, they do add up after a number of tastings.

Dump buckets come in a variety of shapes and sizes. This dump bucket was made from a wine barrel at Mission Estate Winery in Napier, Hawke's Bay, New Zealand.

Cheese/Crackers

Do you think wineries are being stingy when they set out tiny crackers or tiny bites of cheese? Remember, this is not your lunch, and scarfing down the food is not acceptable behavior. The purpose of the crackers, dips or cheese is to cleanse the palate and to help one decide how the wine pairs with food. Some wineries have plates of cheese and charcuterie available for purchase to enjoy while tasting wines. Other wineries have restaurants, so if you want lunch, check out the restaurant.

Picnics

Do you want to have a picnic lunch at the winery? Many wineries encourage visitors to bring a picnic lunch. Frequently, picnic tables or patio tables are available. Ask in advance if it is okay to bring a picnic lunch and where to picnic on the grounds. Do not bring wine from another winery or any other kind of alcohol. Laws restrict wineries, and most wineries are only allowed to have their own alcohol on their premises. Besides, if you were going to someone's home for dinner, it would be tasteless to bring your own entrée.

Politeness

While there are no politeness police around, staff and visitors always appreciate good manners. If you have the opportunity to visit a vineyard, do not pick the grapes. If you would like to taste a wine grape at harvest time, be sure to ask a staff member. One winery we visited mentioned how they do not mind visitors walking through the vineyards and taking pictures, but one day a visitor walked out of the vineyard with several bunches of grapes to ask what they were. It did not dawn on him that he picked the grapes before they were ready. An owner reminded him that the grapes would have made wine if left on the vine. The visitor tried to give them to the owner, who replied, "They aren't good now. They aren't ready for making wine." The best motto to follow in a winery or vineyard would be the National Parks motto, "Take only pictures and leave only footprints."

Helping Yourself

Another major faux pas is to pick up a bottle of wine and pour your own tasting, even if the tasting counter is very busy. Let the wine consultant pour the wine. We did visit a winery in Pennsylvania and one in Colorado that permitted people to pour their own tasting. This practice is very rare. Signs were posted at the two wineries that permitted pouring your own wine.

Buying Wine

When you visit a winery, if you like the wine and want to buy it, that is great, but you do not need to purchase a bottle of wine. Do not buy a bottle of wine unless you like it or can think of a friend who may like the wine. If you do buy wine, make sure you have a proper container to put it in. Placing wine in a car trunk on a hot day may cause heat damage to the wine.

Some people complain that, when they opened a wine at home, it was not as good as it was when they were at the winery. One possible cause is that the wine may have been damaged on the trip home or in storage at their home. Another possibility takes into account the wine tasting event. Tasting wines at a winery is a special event with a unique ambiance for many people. Drinking the wine at home may not be as special. Consider pairing the wine with food and friends. Perhaps light a candle or two.

Tipping

Tipping is not expected at wineries in the United States. However, if you had a particularly good experience and want to leave a tip, you may do so.

Winery Tours

If a tour is available, go ahead and participate in it. Each winery tour focuses on what is important to that winery. Most wineries include a visit to the press pad where the grapes arrive from the vineyard. Also included in the tour may be the tank room and barrel room. Many wineries like to cover the history behind their winery. Some of these histories include winemakers who have come to the

New World to start a winery, and others are entrepreneurs who are fascinated with winemaking. Tours add to the experience and can provide great photo opportunities. Some wineries charge for tours; the fee frequently includes a wine tasting.

Be Prepared

Tasting or drinking wine on an empty stomach is not a good idea. Eat well before you go to the winery. The more food in your stomach, the less the alcohol will affect you. Take a bottle of water and be sure to drink it, especially if you will be visiting another winery. Plan on drinking as much water as you do wine. This is also vitally important when the weather is hot.

More Tips

Many of these tips are common sense. Some other common sense tips include having your ID available, putting your cell phone on vibrate, not chewing gum or smoking while in the winery or tasting room and not complaining about the amount of wine poured for a tasting. Most wineries serve a standard one ounce tasting.

Note Taking

Develop a habit of taking notes while you are tasting wines. There are smart phone apps that you can use, as well as books. You could simply use paper and pencil. The notes can help you decide on particular wines to purchase after a visit.

Keep these tips in mind, and more than likely, you will have an experience you will remember for years to come.

Chapter 3:
Wine Tasting

One of our favorite wine tasting stories came from Rappahannock Cellars in Huntly, Virginia. Five minutes before closing, a loud motorcycle pulled in front of the winery tasting room. A large leather-clad man dismounted a Harley Davidson and dusted himself off. Chains were swaying as he spotted the entrance to the winery and tasting room. Inside, the tasting room manager swallowed and timidly asked, "Would you like to taste a wine?" The biker made a simple response, "Yes." Deciding to skip the whites, the tasting room manager poured a red wine. The biker took the wine and gulped it down in one swallow. Taking a leap of faith, the tasting room manager asked, "Would you like me to show you another way to taste a wine?"

Receiving affirmation, the process of a structured tasting was explained. During each step, the biker patiently followed instructions. Several minutes later, although it must have felt like hours to the tasting room manager, the biker was on his way. However, he suddenly stopped and turned around and said, "Thank you for showing me that. No one ever taught me how to taste a wine."

We, too, learned how to taste wine later in life, although we had instruction from our daughter shortly after she returned from France for a semester of study abroad. It did cross our minds about how she learned her wine tasting skills, and we discovered she took a wine course while in Aix-en-Provence.

Tasting wines is a personal experience and each person's tasting evolves over time. There are a few steps for you to follow, each step designed for you to evaluate the wine. Your evaluation can be as simple as, "I like it" or "I don't like it." On the other hand, your evaluation can become quite complex with vocabulary words that are not familiar to many people. For those living in the United States, the word "cassis" is one of those terms. Chances are, if you ask your friends when was the last time they enjoyed this fruit, you will notice many blank stares.

Some individuals approach the tasting just wanting to know if they like a particular wine from a winery's list of wines. They take the wine and drink it. However, there are steps to a structured tasting that can be followed in a short amount of time. A structured tasting will aid in evaluating a wine. There is nothing wrong, though, with just drinking a sample of a wine and noting if you like it. If you want to apply a more cognitive evaluation, follow the steps of a structured tasting.

Structured Tasting

There are four steps to a structured tasting. Each requires you to use your senses to evaluate the wine. Remember what you taste; whether you like it or not is a personal observation.

Observe the wine in the glass.

Structured Tasting Steps

1. Observe the appearance of the wine.
2. Smell the wine.
3. Taste the wine.
4. Make a conclusion about the wine.

Observe

The first step is to observe the appearance of a wine. Hold the wine glass against a bright white background and tilt the glass to a 30° to 45° angle. Note if the wine is clear or hazy. Is it light or dark? Is it translucent; can you read text through it, or is it opaque and you cannot read text through it? What color is it? White wines can be almost as clear as water, straw colored, shades of yellow, gold or amber. Rosé wines can be pink to orange in color. Red wines can be purple, ruby or garnet. Also, note the effervescence, especially in sparkling wines. Is there just a hint of effervescence on the tip of the tongue, or are there many bubbles, enough that the wine seems to explode in your mouth?

Swirl the wine in a glass to release aromatics.

Smell

Smell the wine in the glass.

The second step is to smell the wine. Some people divide this step into two parts. The first smell is to note what you smell right after the wine is poured into your glass. For the second smell, swirl the wine in your glass. This adds oxygen to the wine and releases aromatics. How intense is the odor in your glass: light, medium or pronounced? Note the aromatics you get from the wine on the second smell. Are they different from the first smell or the same? Aromatics can be divided into broad categories of fruit, floral, spice, vegetal and oak.

In "wine" language, you will frequently hear people use the term bouquet. We prefer the use of the word "aroma" as it seems to be more consumer friendly. However, there is a difference. Aroma generally refers to the fragrant perception of the grape. Bouquet refers to the fragrant perception of the wine including those aromatics picked up as a result of fermenting, maturation and aging in addition to the aromatics from the grape.

Taste

The third step is to taste the wine and note what the taste reminds you of. This step can also be divided into several parts. For the first taste, note the initial shock on your palate. Note if the wine is dry, sweet or in-between. Also, note if the wine's acidity is low, medium or high.

For the second taste, take a small amount of wine and swish it around your mouth, noting the body and what tastes it reminds you of. The body can be light (think about skim milk), medium (think about regular milk) or full (think about cream). Wine tastes can be divided into broad categories of fruit, floral, spice, vegetal and oak. For example, if you pick up a fruit characteristic, try to identify which fruit or fruits. Also, note the tannins that can be non-existent, light, medium or bold. If you are not familiar with tannins, think of a strong cup of tea and how it tends to cling to your teeth.

The third taste is the aftertaste. What do you notice left in your mouth after you tasted the wine? How long does the wine's taste linger in your mouth? This length can be short, medium or long.

Conclusion

The fourth step is to make a conclusion about the wine. Would you say that, for you, the wine's quality is poor, acceptable, good or outstanding? Would you buy this wine for yourself? Would you buy this wine as a gift for a friend?

You can learn the steps of a structured tasting over time. It is important to note that tasting a wine is a very personal activity that recalls information that you have experienced in the past. Terry never drank coffee. When there is a wine that some may say has a coffee taste, Terry does not recognize the coffee notes. There are no right or wrong answers since your notes relate to you.

The more wines you taste using the structured tasting approach, the better you will get at noticing certain aromatics. At the end of this chapter there is a chart that breaks down the broad categories of fruit, floral, spice, vegetal and oak. A mental activity is for you to smell and taste a wine. If you say the wine is floral, ask yourself what flowers. Does the floral component remind you of honeysuckle, daises, roses or some other flowers? You can do this mental exercise for each of the broad categories. With practice, you will discover that you will be able to verbalize different aromatics found in the wine.

Can a Bad Wine Be Good?

Occasionally, people are overheard making comments about wine such as, "I tasted the wines, they were wretched," and "The wine is bad, it's not like west coast wines." Unfortunately, some people may believe that statements like these indicate that the wine is indeed bad. However, these statements tend to tell us more about the individual who made the statement than the quality of the wine. It is easy to classify anyone who knows something about wine as a wine snob. The above wine comments border more on wine ignorance than snobbery.

If someone were to tell you that a wine was wretched or bad, ask him or her to explain what he or she means. Perhaps the wine has a fault, which indeed could make it wretched or bad. Or perhaps the individual simply does not like the wine and refers to things he or she does not like as wretched or bad. To say a wine is bad because it is not like wines from another region indicates a lack of understanding.

Cabernet Sauvignon wines made on the East Coast of the United States are going to be different than the Cabernet Sauvignons made on the West Coast. The Cabernet Sauvignons on the East Coast are generally lighter bodied, more translucent in color, more acidic and have lower alcohol. This does not make them bad; it makes them different. The terroir of the two areas is completely different, and one should expect the wines to also be different. People may certainly prefer one over the other; however, that does not indicate that a wine is bad. A problem arises when one person's opinion influences another person's choice. As you taste wines, know that it is fine if you do not like a wine; someone else may like it.

Wine Faults

If personal preference cannot determine if a wine is good or bad, when can a wine be bad? If the wine has a fault, it may be bad. The problem here is that some people can pick out a fault right away while others may not notice it and may go ahead and enjoy the wine. A good example of this is a wine that is corked. A corked wine has been affected by tricholoro-anisole (TCA) and will smell moldy. Think of the smell of cardboard boxes that have been left in a damp crawlspace for a decade or more. While attending a class at the New York Wine and Culinary Center in Canandaigua, New York, a bottle of wine was opened that was undoubtedly corked. We could identify the moldy aroma two feet away from the opened bottle. The wine did not need to be poured into a glass to pick up the aroma. This wine was bad or even wretched by wine fault standards. It was never served because the presenter immediately knew that it had a fault.

However, a wine that is corked may not always be able to be identified by wine tasters. During a wine tasting at a winery, a corked bottle of wine was served. The moldy aroma was much more difficult to pick up. Many people did not identify the fault. The winery owner, however, picked up on the fault immediately and seized the moment to teach the staff about TCA. One visitor stated that he had tasted the wine from the corked bottle and liked it. TCA can also be present in old barrels and even the wall and wood in a winery. This can also damage wine as the TCA may come in contact with the wine.

At a home wine tasting, our daughter and son noticed an unpleasant aroma from one of the bottles. They opened a wine fault kit and identified the wine as having a smell of glue used for building models. This glue smell indicates the presence of volatile acidity. As with corked wines, some may notice the smell, and others may not. However, this is a wine fault and if the levels of volatile acidity are high enough, the wine will be offensive and should not be served.

Le Nez Du Vin sells the wine fault kit Les Défauts Le Nez Du Vin. The kit consists of twelve vials of liquid that represent twelve faults of wine. These include vegetal, rotten apple, vinegar, glue, soap, sulfur, rotten egg, onion, cauliflower, horse, moldy-earth and cork (TCA). The kit is useful for anyone interested in learning more about wine and what aromas indicate a fault in the wine.

There is debate about whether a wine fault is good or bad. Brettanomyces is a yeast that may affect a wine. People have described the aroma as manure, barnyard and stable. To many, a wine affected by this Brett is bad. However, there are those who believe that a low level of Brett is a good thing. It can add complexity to a red wine. As you taste more wines, look for wines that may be faulted.

Where Do These Wine Aromas and Tastes Come From?

You are tasting an Italian red wine and notice the wine reminds you of cherries. You taste it again and start to pick up licorice. Where did the cherry and licorice tastes and/or aromas come from? Did the winemaker add cherry juice and licorice to the wine? This is a common question. There are so many aromas and tastes associated with wines. Where do they all come from? The answer begins with fermentation.

During the fermentation process, yeast converts sugar to alcohol. Wine grapes contain chemical compounds that are also found in other fruits and vegetables. During the fermentation process, these compounds are released. The cherry aroma and flavor from the Sangiovese grape come from the same compound found in cherries. Different yeasts used in fermentation of wine can create different flavors and aromas. The winemaker has to figure out what yeast should be used based on how the winemaker wants the wine to turn out.

If you ever had a Chardonnay wine that reminded you of butter, the wine underwent malolactic fermentation. In this fermentation, malic acid is converted to lactic acid. A chemical byproduct of this fermentation smells and tastes like butter.

Some winemakers will leave the fermented wines on the dead yeast cells, called lees. This practice creates compounds that will remind you of fresh baked breads and pastries. Oak barrels can add flavors and aromas to wines. Aromas and flavors of baking spices, chocolate, coffee, licorice, vanilla and many more are a result of aging a wine in wood.

Winemaking can become quite complex. Many things done to the grapes, from fermentation on, impact aromas and flavors.

Wine Aroma Charts

The aroma charts, on the next three pages, will give you an idea of just a few of the aromas to look for while tasting wines. Aromas are divided into the broad categories of fruit, floral, spices, vegetal, and oak. Some of these categories are further divided into sub-categories. The sub-categories are then divided into examples. The third chart lists wine faults and examples.

Wine Aromas

Category	Sub-category	Examples
Fruit	Citrus	Grapefruit Lemon
	Berry	Blackberry Raspberry Strawberry Cassis
	Tree fruit	Cherry Apricot Peach Apple
	Tropical fruit	Lychee Pineapple Melon Banana
	Dried fruit/ Cooked fruit	Strawberry jelly Raisin Prune Fig
Floral		Geranium Violet Rose Orange blossom Honeysuckle
Spices		Licorice/Anise Black pepper Cloves

Wine Aromas

Category	Sub-category	Examples
Vegetal	Fresh	Cut green grass Bell pepper Eucalyptus Mint
	Canned/Cooked	Green bean Asparagus Green olive Black olive Artichoke
	Dried	Hay/Straw Tea
Oak	Burned	Smokey Burnt toast Coffee
	Phenolic	Medicinal Bacon
	Resinous	Tobacco Oak Cedar Vanilla

Wine Aromas Indicating Faults

Category	Sub-category	Examples
Oxygen	Acetaldehyde	Rotten apple
	Acetic acid	Vinegar
	Ethyl acetate	Nail polish remover Glue
Sulphur	Sulfur Dioxide (SO_2)	Burnt match
	Hydrogen Sulfide (H_2S)	Rotten eggs
	Ethane Thiol (Ethyl Mercaptan)	Cooked cabbage Garlic/Onion Burnt rubber
	Dimethyl sulphide	Cauliflower
Phenols	4-ethyl-phenol	Horse manure
Fungi	2-ethyl-fenchol	Moldy/earthy Beets
	2,4,6 trichloroanisole TCA	Moldy, musty damp cellar

Wine Flights

Wine tasting at most wineries involve sampling wines from the winery's portfolio. The winery staff will commonly start with the whites and move on to the reds. Tasting is usually done with one wine at a time. Some wineries and wine centers may offer flights. Flights are usually a number of wines grouped by theme, for example a dry wine flight or sweet wine flight. One of the purposes of a flight is to compare and contrast the wines in a flight.

There are different ways to present a flight. We like those winery experiences that have a placemat for four or five wines. Each of the wines are placed on the placemat and there is usually some information about the wine written below it. The advantage of a flight done in this manner is that one can taste all the wines, then go back and retaste the wines. Tasters can also compare and contrast. Some wineries present their flights one glass at a time. This makes comparing and contrasting a challenge since you cannot go back and retaste a wine.

A wine flight should allow you to compare and contrast different wines.

Degustation

We have not seen this term used in the United States; however, in Europe it is much more common. On a trip to wineries in the Istria area of Croatia and the Friuli–Venezia Giulia region of North Eastern Italy, we experienced many wine tastings that were accompanied with food. Degustation refers to the tasting of small portions of foods that focus on the senses. Wines are paired with the various foods. Since, to many, wine is a food, the pairing of wine and food is a great activity that may be offered at some winery tasting rooms.

There are wineries in the United States that also pair wine with food. But we have not seen the use of the term degustation. One memorable pairing was at Vintage Ridge Vineyards in Delaplane, Virginia. The wines were paired with a food tray that included different crackers and bread, cheeses, olives, sausage, cheesecake and preserves, chutney and mustard.

Many European wineries provide for a degustation that pairs wine with food. Breads and food platter at Degrassi Winery in Savudrija, Croatia

Stemware

Stemware is stemware, right? No; not all stemware is created equal. Differences can be seen throughout the wine industry. It may help to know the parts of a wine glass. The top of the glass is the rim/lip. The thinner the rim, the smoother the wine flows into the mouth. The widest part of the glass is the bowl. When pouring a glass of wine, the bowl is filled to only about the 1/3 mark, thus providing room to swirl the wine. Below the bowl is the button where the bowl is attached to the stem and the stem rests on the foot/base.

To some consumers, the stemware a wine is tasted from does not matter. However, we find that it does matter a lot. In Europe, we discovered that wine was usually tasted from fine stemware. Manufacturers such as Riedel, Bormioli, Schott and Spiegelau produce the stemware used in many of the European tasting rooms. We also discovered that many of the wineries in Napa and Sonoma used fine stemware. On a spring trip to Oregon, we observed Riedel stemware in 75% of the tasting rooms we visited. Riedel created a special glass for Oregon Pinot Noir, and the tasting rooms were eager to show off their Pinots in a specially designed glass. Many of the tasting rooms sold the Riedel Oregon Pinot Noir glasses, and they were flying off the shelves. Many wineries in the United States use thick durable small wine tasting glasses. Does the stemware make any difference?

Try tasting water from a styrofoam cup, then a plastic cup, then glass and finally from Waterford crystal. Do you notice a difference? When it comes to wine, the first difference in tasting is noticing the color of the wine. Fine wine stemware has a bowl that is larger and the glass is thinner than many lower-cost glasses used for wine tasting. The visibility of the wine is better. The first smell is different in finer stemware than in small thick glass stemware because there is more room for your nose to smell the wine. Finer stemware glasses are designed to take better advantage of a wine's aromatics. Swirling the wine in better stemware allows more air to mix with the wine and provides a better second nose experience. Then comes the taste.

Choose for yourself. What type vessel would you rather taste the wine from?

We personally like the feel of a thin-bodied, well balanced wine glass. Fine wine stemware has different sizes of bowls designed to place the wine in you mouth at different points. They can take advantage of the wine and the taste buds that will pick up the nuances of the wine. Thick, durable stemware places the wine at the tip of your tongue and keeps your mouth relatively closed. We believe that finer stemware gives one a better opportunity to taste the wine. It will also enhance any wine faults that hopefully will not be present.

Fine stemware is also used in several winery tasting rooms in Niagara-on-the-Lake, Ontario. One winery demonstrates using different Riedel stemware for the different wines we tasted. They also use a thick durable, wine glass. You are encouraged to pour some of the wine from the fine stemware into the thick, durable wine glass and then taste the same wine from both glasses and compare. When side by side, one can tell a difference in aroma and taste.

Since we like to use fine stemware to taste wines, imagine our disappointment when we encountered wineries that used one ounce transparent plastic cups for the wine tasting. These small plastic cups limit your ability to see the true color of the wine. Swirling the wine is virtually impossible. These tiny plastic cups limit two important steps in the tasting experience, the sight and smell of a wine. One winery defended their position by stating that people who come to visit only want to drink not taste their wines. This winery is passing up a golden opportunity to teach visitors how to taste wine.

We attended two presentations of Riedel stemware at conferences in 2012. Each presentation asked the attendees to first taste a wine from a "joker" glass, a glass the hotel uses for their wine tastings. The attendees were then asked to taste the same wine in the appropriate Riedel stemware. We were instructed to note the differences, if any. At both of these presentations, people discovered that there really is a difference. The Riedel tasting also looked at using a wrong glass for a wine. For example, we tasted a Pinot Noir in a Riedel Chardonnay glass and compared it to the same wine in a Riedel Pinot Noir glass. People noticed the differences and preferred the Pinot Noir glass.

Riedel wine tasting compares wine tasted from a joker glass and a Riedel glass.

First impressions make a difference. When we see fine stemware for tasting wines in a winery tasting room, a biased thought is that these wines must be good. On the other hand, when one-ounce plastic cups are introduced the impression is, "what are they trying to hide?" We understand that there is an added cost to using finer stemware for wine tastings. However, finer stemware enhances the tasting experience. More Americans are drinking wine today, and many are more knowledgeable about wine than people were decades ago. The use of finer stemware may keep today's wine tasters lifelong wine enthusiasts.

Evaluation Forms

There are numerous wine evaluation forms and apps that are free or for purchase. They all help you evaluate a wine. Some of these forms suggest you assign a number for many of the evaluation criteria. You add up the numbers, and the sum is a numerical evaluation of the wine.

We have included a very simple form. This form is geared to the beginning wine taster. Experienced tasters have developed their own system. Perhaps the most important evaluation criteria is one that you can do mentally. Simply ask yourself if you like the wine well enough to purchase it. You do not need a form or app to help with this evaluation.

Tasting Form

Winery ______________________
Wine ________________________
Year __________ Price ____________

Step One: Look
Circle the word that most closely defines the color of the wine.
White wines: yellow-green, straw, gold, yellow-brown, amber
Rosé wines: lavender, rose, salmon, orange
Red wines: purple, ruby, deep-red, brown

Step Two: Smell
Circle the word that most closely defines the smell of the wine once it is poured into the glass:
fruit, floral, spice, vegetable, oak, other ________________

Swirl the wine in the glass. Circle the word that most closely defines the smell of the wine:
fruit, floral, spice, vegetable, oak, other ________________

Step Three: Taste
Circle the word that most closely defines the sweetness of the wine: dry, off-dry, medium, sweet

Circle the word that most closely defines the flavor of the wine:
fruit, floral, spice, vegetable, oak, other _________________

Circle the word that best describes how long the aftertaste lasts:
Short, medium, long

Conclusion
Think about purchasing this wine. Circle the word that best describes its quality: poor, acceptable, good, outstanding

Chapter 4: Common International Grape Varieties

As you travel from winery to winery and region to region, you are likely to come across many wines crafted from a variety of wine grapes. While traveling to tasting rooms during harvest, you might have a chance to taste some of these grapes. A few wineries will place a dish of recently harvested grapes on the tasting room counter for visitors to taste. It is interesting to experience the similarities between wine grapes and the wine produced with them.

After tasting wine grapes, you may find yourself dissatisfied with the table grapes that many grocery stores have for sale. Wine grapes are sweeter and have so many more nuances. After tasting wine grapes, you may wonder why grocery stores do not have wine grapes available. Compared to table grapes, the skins on wine grapes are thin, and the grapes do not transport well. Another reason you will not find wine grapes on grocery store shelves is that the price of the wine grapes is much higher than table grapes.

Wines definitely reflect the terroir where they are grown. For example, the Chardonnay grape in some cooler regions offers apple notes, and the wines made from those cool climate Chardonnay grapes will have apple notes. Chardonnay grown in warmer climates may have tropical notes, and the wines made from those warmer climate Chardonnay grapes will also exhibit tropical notes.

As you travel to different wine regions, you will notice the indigenous grapes of the region. These grapes are common to that region and may have limited plantings in other regions around the world. For example, Vernaccia di San Gimignano is common in the Toscana town of San Gimignano, but the grape has little planting elsewhere.

You will also notice in your travels many international grapes. These grapes make excellent wines and are planted in many of the world's wine regions. You will see them over and over while on your wine journey. These are some of the varieties that are good to recognize. This chapter will provide some background information on ten of the international grapes.

International White Wine Grape Varieties

Chardonnay
Gewürztraminer
Pinot Gris (Pinot Grigio)
Riesling
Sauvignon Blanc

International Red Wine Grape Varieties

Cabernet Sauvignon
Garnacha (Grenache)
Merlot
Pinot Noir
Syrah

International White Wine Grapes

Chardonnay

For many years, Chardonnay has been a popular wine for many people. Chardonnay is considered to be an international varietal grape. The grapevine dates back to the Middle Ages in a French village. Ancestors of Chardonnay include Gouais Blanc and a type of Pinot Noir.

Chardonnay is produced in three styles: oaked, unoaked and sparkling wine. This varietal grape is one of the primary grapes used in producing Champagne in the Champagne region of France. Some wine drinkers believe that some Chardonnays are over oaked, while others enjoy the intensity of a strongly oaked Chardonnay.

When tasting a Chardonnay, first determine whether it has been aged in oak or in stainless steel. Then look for nuances in the wine including many types of fruit. Tasters may also note a buttery essence to the wine. Flavors of Chardonnay are influenced and dependent upon the climate where the grapes are grown. Chardonnay is a food friendly wine and pairs well with fruits, fish, poultry and cheese.

Variety	Origin	Notes	Growing Regions	Uses
Chardonnay	France in the Middle Ages	Apple Pear Citrus Peach Apricot Melon Banana Pineapple Toast Vanilla Butter Minerality	Argentina Australia Canada Chile France New Zealand South Africa United States	Salmon Crab Halibut Cheese Chicken Risotto

Gewürztraminer

Gewürztraminer, "spicy Traminer," a white wine grape, produces a very aromatic wine. Some authorities believe that Gewürztraminer, sometimes referred to as Traminer, originated as a mutant of Rose Savigin. The Traminer family has developed into several varieties, with Gewürztraminer being one of the most notable.

Gewürztraminer has been traced back to an early ancestor, Traminer, which likely developed in Italy. Today, many believe that Germany is where Gewürztraminer originated. Be sure to try several types of Gewürztraminer wines. Styles include sweet, dry, sparkling, late harvest and fortified.

Variety	Origin	Notes	Growing Regions	Uses
Gewürztraminer	Germany Italy	Floral Rose Lavender Cologne Perfume Passion fruit Mango Grapefruit Guava Citrus Spice	Australia Canada France Germany Italy New Zealand United States	Cheese Fish Chicken Fresh fruit Sausage Spicy food

Pinot Gris/Pinot Grigio

Commonly referred to as Pinot Gris, in Italy this wine variety is known as Pinot Grigio. Pinot Gris is related to the red wine grape Pinot Noir, according to researchers. The word Pinot Gris comes from the French. In French, "gris" means gray and pinot means "pine cone."

Pinot Gris dates back to at least the 1200s and today is found around the world. Pinot Gris can be bottled as a single varietal wine or used as a blending wine. Late Harvest Pinot Gris wines are also available.

The Pinot Gris grapevine is interesting. As a result of being related to Pinot Noir, the leaves of the Pinot Gris are similar to the Pinot Noir leaves. The grape clusters of the Pinot Gris are unusual, in that, when ripe the berries in the cluster vary in color.

In the United States, Oregon was the first state where Pinot Gris was planted.

Variety	Origin	Notes	Growing Regions	Uses
Pinot Gris/ Pinot Grigio	France	Pear Apple Stone fruits Lemon Melon Spices Vanilla Toast	Canada France Germany Italy New Zealand United States	Rice entrées Seafood including lobster, mussels, oysters and salmon Pastas (white sauce) Goat cheese Green salads Pork

Riesling

The Riesling grape, also sometimes referred to as White Riesling, is believed to date back to the beginning of the Renaissance and possibly even earlier to the Middle Ages. Most likely, Riesling was developed in Germany and thrived there due to the cool climate. The Riesling grape is related to the grape variety Gouais Blanc.

Riesling nuances are heavily dependent on the region where the grapes are grown. Riesling wines can be dry, sweet or even discovered as a true Ice Wine. For anyone interested in botrytis wines, a Riesling wine can occasionally be found. Some wine enthusiasts seek out Botrytis wines, while other wine connoisseurs may not care for it.

Depending on its style, Riesling wines can pair with spicy dishes, pork and smoked oysters. A sweet Riesling could be a perfect ending to a meal.

Variety	Origin	Notes	Growing Regions	Uses
Riesling	Germany	Rose Violet Honeysuckle Citrus Apple Pear Peach Apricot Mineral Petrol	Australia Canada Germany France New Zealand South Africa United States	Aperitif Appetizers Spicy foods Pork Seafood
Late Harvest Riesling/ Icewine	Germany	Peach Apricot Pineapple Mango	Germany Canada	Dessert

Sauvignon Blanc

Sauvignon Blanc is a widely grown grape. With modern technology including DNA testing, researchers have discovered that Sauvignon Blanc is related to Cabernet Sauvignon. The grape is one of the parents of Cabernet Sauvignon. When tasting this white wine, look for mineral and citrus notes.

History

Sauvignon Blanc is believed to have originated in France. Today, the grapes can be discovered on the continents of Australia, Europe, South America, North America and Africa.

Sauvignon Blanc wines pair well with many foods including fish, grilled chicken, sushi and salads.

Variety	Origin	Notes	Growing Regions	Uses
Sauvignon Blanc	France	Minerality Grass Lemon grass Grapefruit Lime Melon Gooseberry Apricot Peach Pineapple Bell pepper Asparagus Vanilla	Canada Chile France Italy New Zealand South Africa United States	Seafood/ sushi Salads Chicken Entreés with green herbs

Red Wine Grapes

Cabernet Sauvignon

In comparison to many other wine grape varieties, Cabernet Sauvignon is one of the relatively new kids on the block. References to Cabernet Sauvignon date back to the 1600s and 1700s, although it was given a different name then. In the late 20th century, it was determined that the Cabernet Sauvignon grapevine is the result of a cross of two other grape varieties, Cabernet Blanc and Cabernet Franc. Frequently, Cabernet Sauvignon is produced as a straight varietal wine. However, it is often blended with Bordeaux grapes including Merlot, Cabernet Franc, Malbec and Petit Verdot. Cabernet Sauvignon is also used as one of the varieties in Italy's Super Tuscans.

Cabernet Sauvignon does well in warm areas. If grown in cooler climate areas, the grapes do not ripen well and the wines produced tend to have unpleasant notes. Wines made with Cabernet Sauvignon and aged in oak tend to age well.

A well-made Cabernet Sauvignon is delightful with a family dinner or at an exclusive restaurant.

Variety	Origin	Notes	Growing Regions	Uses
Cabernet Sauvignon	France	Black currant Blackberry Black cherry Bell pepper Asparagus Green olive Mint Cedar Smoke Coffee Vanilla	Australia Argentina Chile France Italy Spain South Africa United States	Beef Lamb Duck

Garnacha (Grenache)

Garnacha, a red grape variety, is one of the most widely planted grape varieties. Garnacha wine is a food friendly wine and can pair with many types of foods. In France, the grape is known as Grenache.

Many types of wine can be produced with the Garnacha grape. Styles include: dry, rosé and fortified. Frequently, the Garnacha is blended with Syrah, Tempranillo or Mourvèdre.

There is a debate regarding where Garnacha originated. Originally, it was thought to have originated in Spain; however, today Italians believe that the grape originated in Italy.

Variety	Origin	Notes	Growing Regions	Uses
Garnacha (Grenache)	Spain Italy	Blackberry Black Currant W. Pepper Allspice Cinnamon Cloves Vanilla Toffee Leather Toast Smoke	Australia France Italy South Africa Spain United States	Steaks Stews Lamb Pork Poultry Salmon

Merlot

Have you ever wondered where the word Merlot came from? It is believed that the name Merlot comes from the French word merle, meaning "blackbird."

The Merlot grape variety has been known as far back as the first century in Bordeaux. In the Bordeaux region, Merlot is frequently used as a blending grape. Biologically, there is a connection between Cabernet Franc, Carmenere and Cabernet Sauvignon.

Merlot is used both as a straight varietal wine and for blending, especially with Cabernet Sauvignon. In Italy, Merlot is one of the grapes used in producing Super Tuscans. The grapes can be found around the world.

Variety	Origin	Notes	Growing Regions	Uses
Merlot	France	Black cherry Plum Currant Violet Rose Spice Bell pepper Green olive Vanilla Chocolate Smoke Toast Earth Coffee Leather Cedar	Argentina Australia Chile France Italy United States	Hamburgers Steak Seafood Sausage Heavy pasta entrées

Pinot Noir

Pinot Noir has a long history and possibly dates back to the first century. This grape is one of the oldest cultivated Vitis vinifera grapes. The name is believed to have come from the French language, pine (Pinot) and black (Noir.)

While Pinot Noir can prove to be a difficult wine grape to cultivate, it can do well in some regions that have cooler temperatures. In France, one of the best areas to discover Pinot Noir vineyards is Burgundy, while in the United States, look to Oregon and some parts of California. Most regions growing Cabernet Sauvignon, a grape variety that needs warm weather, will not grow Pinot Noir, a cool climate varietal grape.

In the United States, Oregon is well known for its Pinot Noir wines. The wine glass company Riedel was convinced by the Oregon wine industry that a special glass should be created for Oregon Pinot Noirs. The result was a stunning tulip shaped wine glass.

Variety	Origin	Notes	Growing Regions	Uses
Pinot Noir	France	Strawberry Raspberry Cherry Mushroom Earth Violet Rose Vegetal Tea Black olive Leather Vanilla Smoke Toast Cedar	Australia Canada Chile France Germany Italy New Zealand United States	Salmon Pork Beef/Veal Duck Mushrooms

Syrah/ Shiraz

France and Australia are well known for growing Syrah/Shiraz grapes. However, Syrah can also be found growing in other warm climate areas. The grape variety dates back to the 1100s in France. Legends abound in the wine world with Syrah. One legend notes that the grape variety may have come from Sicily, and another legend suggests Iran.

Syrah is bottled as a straight varietal and also used in blends, especially with Grenache. Syrah can also be used in producing sparkling wine.

Ancestors of Syrah include Dureza, a lesser known red grape variety, and Mondeuse Blanche, a white insignificant grape. Petite Sirah, another popular variety, is the offspring of Syrah.

Depending on where Syrah is grown and on winemaking techniques, a large number of wine notes will be discovered as one tastes a number of Syrah wines.

Variety	Origin	Notes	Growing Regions	Uses
Syrah Shiraz	France	Blackberry Black currant Plum Grass Dark chocolate Spice Licorice Cloves Vanilla Leather Cherry Earth Pepper Smoke Toast	Argentina Australia Chile France Italy South Africa United States	Grilled meat Pork Fish Poultry Appetizer: olive oil with bread

Aromas and Tastes Charts

Use the charts on the following two pages as a quick reference to some of the aromatics and tastes associated with the ten international grapes included in this chapter. Do not expect that every varietal wine will have all of the items marked on the chart, but the varietal wine should have some of them.

Aromas and Tastes of Common International Grape Varieties

Aromas/Tastes	Chardonnay	Gewürztraminer	Pinot Gris	Riesling	Sauvignon Blanc	Cabernet Sauvignon	Garnacha	Merlot	Pinot Noir	Syrah
Grapefruit		X	X		X					
Lemon	X		X	X	X					
Lime				X	X					
Blackberry						X	X			X
Raspberry							X		X	
Strawberry							X		X	
Currant						X	X	X		X
Cherry						X		X	X	
Apricot	X		X	X	X					
Apple	X		X	X						
Pear	X		X	X						
Peach	X		X	X	X					
Lychee										
Pineapple	X			X	X					
Melon	X		X		X					
Banana	X									
Passion Fruit		X								
Mango		X								
Plum								X		X
Fig										
Honeysuckle				X						
Violet				X				X	X	
Rose		X		X				X	X	
Lavender		X								
Honeysuckle				X						
Spices		X	X					X		X
Chocolate								X		X

Aromas and Tastes of Common International Grape Varieties

Aromas/Tastes	Chardonnay	Gewürztraminer	Pinot Gris	Riesling	Sauvignon Blanc	Cabernet Sauvignon	Garnacha	Merlot	Pinot Noir	Syrah
Licorice/Anise										X
Pepper							X			X
Cloves							X			X
Cut Green Grass					X					X
Bell pepper					X	X		X		
Lemongrass					X					
Mint										
Green bean										
Asparagus					X	X				
Olive						X		X	X	
Mushroom									X	
Earth								X	X	X
Vegetal									X	
Tea									X	
Smoke						X	X	X	X	X
Coffee						X		X		
Medicinal										
Bacon										
Tobacco										
Leather							X	X	X	X
Toast	X		X				X	X	X	X
Cedar						X		X	X	X
Vanilla	X		X		X	X	X	X	X	X
Baked bread	X									
Mineral	X			X	X					
Petrol				X						
Butter	X									

Chapter 5: Some Common Wine Myths and Misconceptions

As with any subject, myths and misconceptions abound regarding wine, including the history of wine, winemaking, vineyards and the beverage itself. Everyone who has tasted wine has an opinion. This chapter takes a look at several misconceptions and myths.

Red wine is from red grapes; white wine is from white grapes.

Not so! The juice from red and white wine grapes is clear. Red wines are usually fermented on their skins after crushing. The red to purple to black color of red wine is picked up from the skins of the grapes. White wines are made by pressing the grapes after harvest. The juice is then fermented. Some winemakers throughout the world are making white wines with some skin contact. This usually results in a dark yellow to dark gold color of the white wine.

Interestingly, we heard from a winemaker from a Canadian winery that, for several months, people would stop by the winery and ask for a red Chardonnay. The winemaker could not figure it out. Comically, he noted that eventually he did discover a winery producing a red Chardonnay, and this was "a smack against common sense." Possible explanation: since Chardonnay is traditionally known as a white grape, one way to produce a red Chardonnay would be to allow the Chardonnay juice to have contact with the skins of a red grape, such as Chambourcin, during fermentation. Adding a different grape's skins to a fermentation is not a new concept. For some time, winemakers have added Viognier skins to Syrah during fermentation. The Viognier skins give the Syrah some floral notes.

Another way to introduce a red color is to blend a bit of red wine with the Chardonnay wine. If you blend a red wine with the Chardonnay wine, can you call it Chardonnay? This goes back to the strict labeling regulations that generally say that, for wine labels to list the name of the grape, the wine must be produced with 75 percent of the designated grape. So, a winemaker could blend red wine to the Chardonnay and still label it Chardonnay if it is below a certain percentage.

Every winery tasting room is noisy.

Wine Trail Traveler says, "Not necessarily; most tasting rooms are not like a bar scene." In general, winery tasting rooms are friendly but more subtle, fun experiences. If you want an especially quiet experience, visit a winery during a weekday or during the offseason. This gives winery staff time to talk with visitors. Crowded weekends during the high visit season do tend to be noisier.

Every wine drinker should enjoy red wines.

Don't believe it. White wines are just as legitimate and tasty as red wines. Rosés and orange wines are perfectly fine. It all depends on the characteristics you are looking for in a wine. In general, we like red wines with dinner and white wines on the patio.

The more expensive a wine, the better it will be.

This depends entirely on the consumer. If you enjoy $15 bottles of wine, you may not enjoy a $150 bottle of wine ten times as much. You may not enjoy it at all. On the other hand, if you start drinking many $150 bottles of wine, you may not enjoy the less expensive wines as much as you once did.

The price of a bottle of wine is determined by many factors. Different grape varieties demand different prices. Growers who sell Niagara and Concord grapes may only receive a few hundred dollars a ton, whereas growers of vinifera grapes can receive over a thousand dollars a ton in some areas. The cost of the grapes is also associated with the location the grapes are grown. Expect to pay more per ton of grapes grown in Napa Valley compared to Lake County just to the north of Napa Valley.

If the wine was aged in barrels, the expense increases. Barrels cost the winery, and this cost is passed onto the price of the wine. American barrels cost a few hundred dollars per barrel, and French barrels can cost over $1,000 per barrel. The length of time aging a wine in a barrel will drive up the cost. Barrels are usually stored in a temperature and humidity controlled room. This increases the cost for a winery and will be passed on to the consumer when determining a price for the wine.

Different closures have different price points, as do bottles. The amount of time the bottled wine is stored at the winery before release to the public will also determine the price of a wine. In addition, there are different winemaking techniques that can drive up the cost of the wine. For example, sorting the grapes on a double sorting table before fermentation is labor intensive and will increase the cost.

Generally, a higher cost bottle of wine probably was made with ultra premium grapes. The wine was aged in expensive barrels for an extended period of time and bottle-aged for a prolonged period of time. This does not mean you will like the more expensive wine better though. Despite how expensive a wine is, the most expensive and well-made bottles of wine subjected to heat damage or perhaps a tainted cork will not be good.

Wines should be aged. Wine gets better with age.

Most white wines and many red wines do not need to be aged. Winemakers will sometimes call these fresh wines, and they are meant for you to drink now. The best red wines to age are those that have good tannin structure. Many consumers buy a bottle of wine and drink it within a day or two.

A wine sealed with a cork means it is a better wine than one sealed with other closures.

Screw caps and synthetic closures are being used more and more. Corks have been known to cause cork taint of a wine, which destroys the wine. Screw caps and synthetic closures are legitimate closures for wine bottles. Other types of closures include the Zork, a plastic spiral-type closure, and glass closures.

The more well known the winery, the better the wine.

This misconception is probably a result of not only good wines but also great marketing skills. Many small wineries craft excellent wines. Since production is low, they may not have national and international distribution. They are gems waiting for your discovery.

Wine is made in the vineyard.

Oops! Exactly what is the role of the winemaker? While a winemaker cannot make a great wine with poor quality grapes, the winemaker has an important role in producing great wines. There are hundreds of decisions that a winemaker makes before you drink the wine. If two winemakers made wine from the same variety of grapes from the same vineyard location, the two wines would likely be different.

We believe that winemakers and viticulturists should take the credit for the wines they produce. Both winemakers and vineyard growers deserve credit for what they do well.

It is easier to grow grapes than it is to make wine.

Not enough credit goes to the hard work that a vineyard manager does and the knowledge he needs to grow quality grapes. Growing quality grapes is only the first step in the process of producing wines. Growing grapes is not easy. Grapevines demand hours of a grower's time throughout the year. Grapevines are not like crops that can be planted in late spring, harvested in late summer or fall, and the rest of the time the farmer has off. With grapevines, growers have to tend the vines 12 months a year. Pruning the vines takes place during the winter when the grower braves the cold to often hand prune the vines.

One needs to be a wine connoisseur to enjoy wine or enjoy visiting a tasting room.

Anyone 21 or older can feel comfortable in a tasting room. The winery staff is friendly and many enjoy telling their stories about wine or explaining the history of the winery and talking about the vineyards and winemaking process. If you do not know much about wine, visiting tasting rooms can provide a gentle wine education. Wine connoisseurs can also enjoy visiting tasting rooms since they are the only place that one can sample a winery's portfolio of wines.

One does not need to drink water when tasting wines.

Those small tastings of wine can add up quickly; be sure to drink water while tasting wines. A good rule to follow is to drink as much water as wine.

When buying wine, I should listen to the wine critics.

Wine critics know about wine. But, and this is important, unless your tastes are exactly the same as a particular critic, you may not enjoy what they like. Tasting is based not only on your tastebuds but also on the experiences you have had with foods since you were very young. Although a critic can provide one piece of data, decide for yourself what you like and do not like.

Pair red wine with meat and white wine with fish.

This myth is quickly earning its demise. Not everyone enjoys a red wine, so why would you want to pair red wine with anything? Today, many people are enjoying red wines paired with fish. There is also a tendency to drink what you like regardless of what food you are having. On the other hand, sometimes a wine can make a food better, and a food can make a wine better. Discover for yourself these magical combinations. Be open to pairing suggestions, and then determine if you like them.

Icewines must be kept in the refrigerator.

No, unless the bottle has been opened. Be sure to use it quickly as wines tend to deteriorate after they have been opened. Unopened bottles of Icewines can be stored in a wine room or cellar.

Legs determine the quality of the wine.

Legs, sometimes referred to as tears, appear on the sides of a wine glass after the wine is swirled. One of the causes of "legs" is the alcohol content in the wine. Legs do not provide an indication of the quality of the wine.

I can hold the wine glass anyway I want.

Sure, go ahead. Hold the glass by the bowl. It's a great way to look like a newbie to others. Part of this "hold it by the bowl" tendency may be attributed to the numerous television shows and movies where actors frequently try to look suave by holding stemware incorrectly. Where are the wine critics who allow that a free pass? By the way, you are not a "wine snob" if you hold a wine glass by the stem the way it was intended; you just happen to know something about wine. Holding by the bowl may cause the temperature of the wine to rise. Assuming the wine was served at the correct temperature, one should avoid holding the stemware by the bowl.

I don't like a wine; it must be bad.

Wine tasters should tell staff why they do not like a wine. Saying a wine is bad does not help anyone. A description of why you do not like a wine helps tasting room staff members know what type of wines they should present to you. Is it too sweet for your taste? Someone else might like that sweetness. Knowing a few wine faults and saying a wine has a fault is helpful. One of the most common wine faults is corked wine which may be caused by a cork or contamination in the winery.

In a tasting room, if you do not like a wine, quietly explain to the staff why you do not like a wine. This will help him/her guide you to other wines you might like better.

I hear they can't make good red wines in XXX (insert any location.)

This statement is just so misleading. What is the definition of "good?" Do not use the words good or bad without describing why. Just because you like a wine does not mean someone else will like it. We, while frequently agreeing on what wines we like, also disagree on some wines. Do not denounce an entire wine region. It is unfair to the regional winemakers and other wine enthusiasts who might enjoy wines from that region.

Refrain from judging a wine region on hearsay. Travel to the area and discover the wines. You will be in a much better position to determine if you like the wines from a particular region.

Red wines should always be decanted.

No, young wines and red wines with few tannins do not improve with decanting. Some red wines are not filtered and there may be a bit of sediment at the bottom of the bottle. This does not indicate any problem with the wine. If a wine is decanted, it helps to keep the sediment from being poured into the glass. Either way the wine is fine.

Rinsing with water between tastings is a good idea.

According to winemaker and winery owner Chris Pearmund (Pearmund Cellars, Broad Run, Virginia), "Rinsing with water between tastings, many people think it is a good idea, but wrong they are." While rinsing a wine glass with water removes the telltale signs of the previous wine, the water drops dilute the next wine. One alternative is to rinse the glass with the next wine and then refill. While at a sommelier conference, everyone did rinse with water. They then turned their glasses upside down to drain on a napkins for several minutes. A few upscale wineries will serve tastings in separate glasses; however, expect to pay more for the tasting.

A winery in Mendocino County, California, noted that people do not seem to understand the labeling of organic wines.

In the United States, the TTB (Alcohol and Tobacco Tax and Trade Bureau) approves wine bottle labels. Recently, regulations regarding the use of the word "organic" on labels changed. If a wine bottle says "organic wine," it must be produced with grapes maintained with organic standards and the winery procedures must use organic methods.

Label Designations for Organic Wines

100% Organic – means that the wine has all organic materials with no added chemical sulfites.

Organic – wine is made with a minimum of 95% organic ingredients and no added chemical sulfites.

Made with Organic (Ingredients) – wine is made with a minimum of 70% organic ingredients with no additional chemically added sulfites.

Any wines made with less than 70 percent organic materials may only list the organic materials in a non-conspicuous place like the ingredients statement.

Arnoldo Caprai Winery in Italy sites two common misconceptions, Sagrantino is Sangiovese and Montefalco is in Tuscany.

Wine Enthusiast Magazine awarded Arnoldo Capri Winery the 2012 European Winery of the Year. This winery, in Montefalco, welcomes wine travelers, many of whom have two misconceptions, one dealing with their location. Many visitors believe Montefalco is in Toscana (Tuscany) when it is really in Umbria. The region Umbria does share a border with the region Toscana. The other misconception deals with the grape Sagrantino, perhaps the signature grape of the Umbria region. It is believed by some that Sagrantino is just another name for Sangiovese, the main red grape of Toscana. However, the two grapes are completely different with totally different profiles.

Grapes don't grow in Wales because it is always raining.

Judith Dudley from Parva Farm Vineyard in Wales wrote to say that people think that wine cannot be produced in Wales because it is always raining. We can attest to the misconception of the statement. When we visited Parva Farm Vineyard it was not raining! The vineyard was doing fine. Grapes had been planted in Wales since the time of the Romans.

Chapter 6:
Tips for Planning to Visit a Wine Region

With tens of thousands of wineries and vineyards around the world, deciding on a wine region to visit can be a daunting task. There are over 7,000 wineries in the United States, and every state now has wineries. It would take decades to explore the wine regions of Europe, Africa, South America, North America and Oceania. The planning can be pleasant, though. As you sift through information about wine regions, you can learn about those regions, the grape varieties they grow and the wines they make.

You will have to decide when you want to travel. Each season has its advantages and disadvantages. Are you going to research and develop your own itinerary? This takes more work but can become very personalized. Another option is to contact a wine tour operator that organizes wine tours in the region you want to visit. There are an assortment of wine tour operators. Some will develop an itinerary and contact the wineries while you drive from one winery to another. Other wine tours may include transportation and a guide.

This chapter will provide some background to help you begin planning to visit wine regions.

Best Time to Visit

Many people enjoy visiting wine country during harvest. The grapes hanging on the vines paint a romantic notion of vineyards and forecast the romance of the wine to be crafted. Harvest is not a one-day event. Each grape variety is picked at a different time. Therefore, harvest can last for months if a vineyard has several grape varieties. Late summer and early fall are ideal times for wine tourists to visit winery/vineyard tasting rooms. In the northern hemisphere, harvest can begin in July in the southern regions and last until October further north. If a vineyard is harvesting grapes for a late harvest wine, the grapes may hang on the vines into November. Tasting rooms during harvest can be crowded, especially on weekends. Visiting on weekdays may be enjoyable if you want to learn more about the winery and their wines.

Harvesting grapes at Wooden Valley Winery and Vineyards in Suisun Valley, California

In the fall, vineyards can put on a show of color. October and November can be ideal travel times in the northern hemisphere. Depending on the variety of grapes, vineyards can displays colors of deep reds to oranges to yellows. The height for visits to wine regions is usually from late summer through mid fall. Winemaking takes place during this time. Visitors to wineries may observe crush, press, fermentation, punch down and racking. However, other seasons have their advantages.

Winter visits are peaceful. Vineyards sleep, and wine travelers can observe a vine's architecture. You may also see vineyard workers braving the cold to prune the vines for the upcoming growing season. Snow-covered vineyards and snow clinging to the vines make for ideal photography. Tasting rooms are quieter during the winter. You may have a better chance to speak with the winemaker or owner during the winter. Wine tastings tend to be leisurely, and you can have long conversations. Winery/vineyard tasting rooms are seldom crowded even on weekends.

A spring visit also affords a quieter time. Vineyards are pushing forth new growth, and you may observe small clusters of caps. Later, flowers will open, and the caps fall off. Still later, you will see buckshot berries spread out in a cluster. The winery may have bottled some white wines from the previous year's harvest. The pace is still slow, even on weekends.

During the early summer, you will notice more people in the tasting rooms on weekends; however, visiting wineries during the weekday is often a treat. There are fewer people, and this gives visitors a better opportunity for a leisurely wine tasting and conversation. In the vineyards, there is more growth of the vines. The buckshot berries give way to larger berries that touch each other.

Decide for yourself the best time to visit wine country. Each season is unique and offers different photographic possibilities. There will be crowded times and quiet times for you to consider.

How to Find Wineries to Visit

So, you decided to plan a trip to visit a wine region, and now you need to decide which tasting rooms to visit. Your first tool for research can be the Internet. You can start by conducting an Internet search for wineries in a particular state, province, country or wine region. For example, if you decided to search for "wineries in Virginia," Google and Bing will display hundreds of sites. At the top of the list is the VirginiaWine.org website. This site provides plenty of information about Virginia wineries. Not all states, provinces, countries and wine regions have a site that lists wineries. Another search option is to search for "grape growers or winemakers associations" or "wine trails" in a wine region. A search for Mendocino County grape growers association will yield result pages with a link to the Mendocino Winegrape and Wine Commission at the top of the search results. The website offers a tool for visiting wineries in the region. Wine Trail Traveler (www.winetrailtraveler.com) has 60 suggested winery itineraries to peruse.

Another tactic used to locate wineries is to use Google Maps or Bing Maps. For instance, if you are planning a trip to Montreal, Quebec, and you want to know the wineries near the city, type "Montreal, Quebec" in the search box of either Google Maps or Bing Maps. A map of Montreal will display on the webpage. Next, you can zoom out to cover a region larger than the city. Then, type "wineries" in the search box. Google or Bing will display a number of potential wineries on the map. Not every bubble or circle on the map will be a winery. It will also include wine bars and wine stores. If you use "vineyard" as a search term, the map will also show the locations of churches that have "vineyard" in their name. You do need to do some research if you use this method. Someone who is a visual map kind-of-person will find this a valuable tool to locate wineries in wine areas throughout the world.

You can also search for sites that have mentioned wineries in a region. In the Google or Bing search box, type "best wineries to visit in Champagne." The search will generate millions of webpages for you to sift through.

By using these research methods, you can generate a list of potential wineries to visit. Once a potential list of wineries in an area is established, find out if the wineries are open during the time frame of your visit. Some wineries are open daily throughout the year, while others may only be open to the public on weekends. There are those wineries that close during the winter and those that open only during the summer. It is best to check ahead of time to see if the winery tasting room will be open during the time of your visit. You can often find opening hours on a winery's website. If not listed, call the winery.

After you have narrowed down a list of potential wineries to visit during an upcoming vacation, should you just show up and walk in or should you make an appointment? Although you may enjoy the random act of just walking into a winery tasting room, some preparation should make your visit more enjoyable. Some wineries require appointments, and often these need to be made prior to the day of the visit.

Since many wineries have websites, visitors can check out the website for an idea of what to expect while visiting. Check for the hours of operation, the wine list and information about tasting fees and tours. It is a good idea to know that there is a tasting fee before going to a winery. It is also worth knowing how many wines one can taste. Many wineries limit the number of wines for tasting to five or six. Knowing the wines one wants to taste prior to visiting the tasting room will help you from becoming disappointed. If information about the number of wines and tasting fees is not available on a website, call the winery.

Some people are interested in a more thematic wine holiday. If there are many wineries in a region, you can locate wineries that are grouped by theme. If you are interested in organic and bio-dynamic wine, visit wineries that are organic or bio-dynamic. Perhaps you would enjoy visiting wineries that have a small production. Other themes include: wineries in historic buildings, wineries/vineyards with world recognition or wineries/vineyards with certain classifications such as Grand Cru Classé. Be creative, and think of your own theme.

Driving to Tasting Rooms

It is a good idea to get directions to winery tasting rooms. GPS systems work well but have been known not to recognize an address. Some wineries have a short street or long driveway leading to their parking area. They have named the street or driveway, and occasionally, the GPS will not recognize the name. If this happens, enter the crossroads closest to the winery. While driving to wineries in the Istria region of Croatia, we discovered that sometimes our

GPS shows the exact location of the vineyard on the 45th parallel at 45 North Vineyard and Winery in Lake Leelanau, Michigan.

GPS took the address; other times it took the postal code. Many winery websites have directions to their tasting room. These directions should trump GPS directions. One can also get directions from several online mapping sites such as Google Maps and Bing Maps.

Some wineries are listing their GPS coordinates. These coordinates should lead you to the exact spot the reading was taken, hopefully the parking lot. Be aware there are different formats of writing coordinates, and these formats are not interchangeable. We learned this the hard way. While traveling to wineries in the Lake Erie North Shore wine region of Ontario, we entered the GPS coordinates in the format we were used to. We ended up several kilometers off target. We then learned that our GPS has three different formats that we can use to enter coordinates: h ddd.ddddd° or h ddd° mm.mmm' and h ddd° mm' ss.s." Using the correct format, we were able to locate the winery.

Wine Tour Operators

Arranging an itinerary yourself is time consuming. For a day trip, an itinerary may be easy. If your journey involves multiple days outside of your region, a wine tour operator may appeal to you. Some wine tour operators are small and work only in one wine region. They often will have a car or van and will take small groups. The driver talks about the region and wineries you visit. Usually, the driver will pick you up and return you to your hotel. This is very convenient. You do not need to make an itinerary, and you do not have to worry about driving. You can sit back and relax and enjoy the excursion. Many of these tour operators know the winemakers and owners and, to a certain degree can, customize a trip for you.

We have experienced two small regional tour operators, one in Virginia and the other in California. Since we had visited the wine regions previously, we were a challenge to the tour operators since we had already visited the wineries they usually go to. Neither tour operator was daunted by the challenge, though. They both were able to create an itinerary that included wineries that we had not previously visited. Both tour operators picked us up at our lodging facility, took us to the wineries, provided a lunch and dropped us back at our hotel. The winery visits were relaxing and enjoyable. It was also nice to have a conversation with the tour operators, also wine enthusiasts.

Another wine tour in Virginia was a larger tour. A small bus took several of us to different wineries in Northern Virginia. The company also offers wine tours in Central Virginia. A guide on the bus introduced us to the region and the wineries. We had plenty of time to meet and talk with the other wine tourists on the bus and during our tastings at the wineries. A benefit of a larger group tour is the number of other wine enthusiasts you can talk with.

These regional wine tour operators know the local wineries, winemakers and owners. The smaller your group, the more they can tailor a tour to meet your needs. For international travel, we had different experiences. There are international tour operators who will conduct wine tours to many countries and several wine regions within a country. We have been on a tour of New Zealand on a bus with a large group. We also visited wineries in Croatia by renting a car and driving ourselves.

The bus tour of New Zealand wineries was very relaxing, and over the two-week period, we were able to spend time with the other 26 people on the bus. The bus drivers were very knowledgeable of the areas they were driving and narrated along the way. Our tour leader was a member of our group who helped organize the tour. This was a very easy tour. Awake in the morning, enjoy breakfast, board the bus, head to wineries, have lunch, head to more wineries and return to the hotel. Our evenings were free to explore the neighborhood, eat dinner and enjoy the companionship of the group members. Many of this group were winemakers and wine growers.

Our wineries tour of Croatia was a bit different. This was a custom-made tour for us. Each of the ten wineries we visited in Croatia had a staff member who spoke English, usually the owner or winemaker. A tour company set up our itinerary, procured a rental car, made the reservations for our lodging and made appointments at the wineries. We just had to pick up our car at the Zagreb airport and hit the road with the maps the tour operator provided. We did get lost a couple of times; however, we were using a GPS during those times, so we never stayed lost forever. If you like driving, this tour would be a nice option. We discovered that, when we drive, we have a better idea of the landscape and what is outside our vehicle. When on a bus, we end up talking and missing the landscape.

Terry and Kathy Sullivan are joined with Amandine Vicente Biosca, and Karène Martel-Dumas of In Vino Veritas Travel Agency at Union Libre in Quebec, Canada. Photo by In Vino Veritas Travel Agency

Wine tour operators and their wine tours will cost more than doing all the planning yourself. However, everything from winery visits, lodging and transportation is arranged for you. Your challenge is to keep on schedule and not miss the bus.

The Romance of Visiting Wineries/Vineyards

People like to associate wine with romance. Light a candle and share a bottle of wine, and you will create a romantic setting. Consider visiting wineries and vineyards that offer accommodations on their property. The notion of romance extends to waking up in the morning and seeing vineyards, strolling along the vines and experiencing the local food first hand. Spending the evening at a winery or vineyard often brings the traveler in contact with the owners, winemaker or vineyard manager. There may be an opportunity to participate in the daily tasks taking place in the vineyard or winery. Check with a wine tour operator to see if they customize a tour with onsite lodging. You can also conduct your own research.

Appendix

Wine Tourism Club

Other than a wonderful experience at a winery tasting room, discovering their wines and the romance of vineyards and winery, there should be some benefit to visiting wineries around the world. We created the Wine Tourism Club for wine travelers. Keep track of the winery/vineyard tasting rooms you visit. You can download the application on the following pages from http://winetrailtraveler.com/travel/winetourismclub/winetourismclub.php.

Print the application and take it with you when you visit wine regions. Once you have visited 100 wineries send the application in to Wine Trail Traveler for a certificate and bragging rights. Membership in the club is free and it has several levels.

# wineries visited	Membership Level	Certificate
1 - 99	Pioneer	
100	Pathfinder	X
200	Voyager	X
300	Wayfarer	X
400	Crusader	X
500	Globetrotter	X

Wine Tourism Club Application

First Name:

Last Name:

Email:

Directions:
Add the date and the winery name you visited. You can print this application and keep it in your car or carry with you when you visit wineries.

When finished, send a copy of the application to Wine Trail Traveler LLC by mail or scan the application and send by email.

mail: Wine Trail Traveler LLC
6578 Dovecote Dr.
Columbia, Maryland 21044 United States

email: winetourismclub@winetrailtraveler.com

Wine Tourism Club Application

#	Date	Winery Name
1		
2		
3		
4		
5		
6		
7		
8		
9		
10		
11		
12		
13		
14		
15		
16		
17		
18		
19		
20		
21		
22		
23		
24		
25		
26		
27		

Wine Tourism Club Application

#	Date	Winery Name
28		
29		
30		
31		
32		
33		
34		
35		
36		
37		
38		
39		
40		
41		
42		
43		
44		
45		
46		
47		
48		
49		
50		
51		
52		
53		
54		
55		
56		

Wine Tourism Club Application

#	Date	Winery Name
57		
58		
59		
60		
61		
62		
63		
64		
65		
66		
67		
68		
69		
70		
71		
72		
73		
74		
75		
76		
77		
78		
79		
80		
81		
82		
83		
84		
85		

Wine Tourism Club Application

#	Date	Winery Name
86		
87		
88		
89		
90		
91		
92		
93		
94		
95		
96		
97		
98		
99		
100		

Once you have visited 100 winery tasting rooms, send the application to:

Wine Trail Traveler
6578 Dovecote Dr.
Columbia, MD 21044

or, email the application to:
winetourismclub@winetrailtraveler.com

__ Check here if you would like to receive the free Wine Trail Traveler Quarterly pdf. published in January, April, July and October.

Glossary of Wine Terms

This glossary is a selection of wine related terms which are fun to know but no one should feel it a necessity to know these words in order to enjoy a nice glass of wine.

ABV: Alcohol by volume.
amphora: A clay vessel for storage and transporting wine.
appellation: The legal geographical designation for grapes.
aperitif: A wine that is drunk by itself without food. It also refers to a wine before a meal meant to stimulate the appetite.
aroma: Refers to the fragrant perception of the grapes.
attack: A wine's first impression.
balance: A combination of tannins, fruit, alcohol and acids working together. If one aspect overrides the others, the wine is not balanced.
barrel: A cylindrical container used to age wine. Barrels can be made of wood or stainless steel. Wood barrels are mostly made of oak, however; there are barrels made with acacia wood. Some winemakers prefer stainless steel barrels when producing a wine with lee contact but no influence of wood.
blind tasting: Tasting a wine without knowing what it is. Useful when evaluating a wine or a fun experience at wine parties.

bouquet: Refers to the fragrant perception of the wine including those aromatics picked up as a result of fermenting, maturation and aging in addition to the aromatics from the grape.

brix: Measurement to determine the sugar content. For example 21°B is juice that has a brix measurement of 21 grams of sugar per 100 grams of juice. This measurement is used by growers as data to help determine when to harvest grapes. It is also used by winemakers to determine the potential alcohol for juice and the amount of sugar converted to alcohol.

brut: Refers to a very dry sparkling wine or Champagne.

calcium carbonate: CaCO3 A winemaker will add calcium carbonate to a juice or must if it is too acidic.

carboy: A vessel often made of glass or plastic that is used to age wine. Home winemakers often will use three, five and six gallon carboys for home winemaking.

chaptalization: Adding sugar to wine to increase the residual sugar. This will increase the potential alcohol of the wine.

cold soak: Allowing certain grape varieties to soak in cold temperatures for a number of days prior to fermentation. Winemakers will do this in order to extract more color, aromatics, flavors and tannins.

crush: A process where gentle pressure breaks the skin of a grape and releases the contents.

demijohn: A vessel used for fermenting and aging wine, similiar to a carboy.

destem: Removing the stems from grapes.

fermentation: The process where yeast converts sugar into alcohol.

filter: A winemaker filters a wine to remove particles and help make the wine clearer. Often filtering is done before bottling the wine.

fining: A process whereby a substance is added to the wine to clarify the wine. The substance has particles adhere to it and then falls to the bottom of the container. The result is a clear wine without sediment.

finish: After you swallow or spit the wine, the finish is the last flavor you notice. Writers often describe the finish by using a time period such as " short finish" or a "long finish." After swallowing or spitting, the flavor remains for a short time or a longer time. A long finish is often associated with a quality wine.

fruit wines: a wine produced from fruit such as: apples, blackberries, blueberries, raspberries, strawberries and more. Some of these fruit wines are produced from 100% fruit while others use a grape wine for a base and add a fruit flavor to the base wine.

gyropalette: A large crate holding bottles of sparkling wine or Champagne. This crate turns to allow riddling to take place.

harvest: The time of year when the grapes are picked by hand or machine. In the northern hemisphere this is typically between July and November. Harvest occurs in the southern hemisphere between February and April.

Icewine: Also Eiswein in Germany, a dessert wine made with grapes harvested when temperatures drop below 17°F. Grapes are crushed at this low temperature. There are only a few drops of juice per grape. The juice has a high concentration of sugar.

lees: Dead yeast cells that build up the sediment at the bottom of a barrel or fermentation container.

maceration: A winemaking process where grape skins are combined with the juice. This gives the color to red wine and intensifies the flavor.

malolactic fermentation: A secondary fermentation whereby malic acid, which is harsh, is converted to lactic acid which is softer.

MOG: An acronym for "materials other than grapes."

must: A thick mixture of skins, pulp, seeds and juice.

oenology: Also enology, the study of winemaking.

pH: A measurement of a solutions acidity. The scale is logarithmic and the lower the measurement the higher the acid. A pH of 3.0 is ten times more acidic than a pH of 4.0. Bottled wines with a pH of 4.0 and higher are suspectable to spoilage. Many finished wines are near the pH 3.5 range. Note that the higher the sugar content of a wine the lower the pH needs to measure (lower pH means more acidic). This ensures that the wine will have a crisp finish and will not taste like syrup. Winemakers can adjust a wines pH usually by adding tartaric acid.

phylloxera: A grapevine disease caused by small aphids that destroy grapevines.

pomace: The grape skins, seeds and jacks after the grapes are pressed.

potassium metabisulfite: A substance used by winemakers to inhibit bacteria growth. It is also used in higher doses to sanitize equipment.

press: A winemaking process where the winemaker uses a wine press to extract the juice or wine from the grapes or must.

qvevri: Also kvevri, a clay vessel of different sizes. Qvevri are used for fermentation and aging wines. Qvevri are buried underground up to their necks. This helps maintain a constant temperature. Qvivri have been used for winemaking for 8,000 years making them the oldest winemaking equipment.

racked:Wine that has been moved from one container (barrel, stainless steel tank, carboy) to another. Winemakers rack a wine to eliminate the solid particles that have fallen to the bottom of the container during fermentation and aging. Once the wine is racked the original containers are cleaned and reused.

refractometer: Instrument used to determine the sugar level in grapes, some are small enough to take to the vineyard to analyze the sugar level. The reading is in degree brix. The winemaker can convert the brix into potential alcohol. A reading of 20 brix will create a wine around 11% alcohol. A quick way to do the math mentally is to multiply the brix by 0.55. This will give an approximate potential alcohol level.

riddling: Also known as remuage, riddling is the process of tapping and turning a bottle of sparkling wine downward increasing the angle of the bottle to 90° so that the bottle cap is pointing downward. This process will cause the lees in the bottle to settle in the bottle's neck next to the cap. The process can take several weeks.

sommelier: Wine consultant in a restaurant.

stave: The oak pieces of wood that are put together and formed to make an oak barrel.

tank: A container used for fermenting and aging wines. Tanks have various sizes and are constructed of wood, plastic or stainless steel.

TA (Total Acidity): Is a measure of a wine's total acidity. The range is 0.0% to 1.0%. A desirable result is 0.6% for red wines and a bit higher for white wines. As a wine's TA nears 1.0% the wine becomes sour and too tart for most people. As the TA nears 0.0% the wine will taste flat. Wines with low TA are susceptible to spoilage.

tannins: Compounds found in grape seeds and skins that cause the mouth to pucker.

terroir: The effect of the combination of soil, weather, and climate on grapes in a given area.

thief: Instrument used to siphon wine from a barrel, carboy or any wine container for a tasting.

tierce: One-third of a barrel.

toast: Similiar to how a toaster will cause bread to toast, building a fire inside of a barrel will cause the wood to toast. Just as in bread, the degree of toast is controlled.

vat: A container for fermenting and aging wine. In France, many vats are made of concrete. There are different sizes often corresponding to the grapes from a particular block in a vineyard.

veraison: When grapes are just beginning to show signs of turning color.

vertical tasting: Tasting several wines from different vintages. Useful when comparing and contrasting the different vintages.

vintage: The year the grapes are harvested.

vintner: A individual who makes wine.

viticulture: The study of growing grapes.

weeping: In a vineyard just before budbreak, a small drop of clear liquid will form where the grapevine was prunned.

wine enthusiast: One who enjoys wine and is eager to learn more about it.

wine snob: One who believes he or she has a perfect palate and knows "everything" about wine.

yeast (commercial): Hundreds of different yeasts are available for winemakers to choose from.

yeast (natural): Yeast naturally occurs in the vineyard and some winemakers rely on naturally occurring yeast to use in fermenting their grapes.

Made in United States
Orlando, FL
17 June 2026

82994898R00056